STEVE PARISH

Discover & learn about
Australian
Deserts
AND ARIDLANDS

Text Pat Slater

Photographs Steve Parish

HABITATS & ECOSYSTEMS

Contents

See page 16.

See page 12.

See page 8.

INTRODUCTION

The voyage of Ark Australia 3

ABOUT AUSTRALIA'S DESERTS

What is a desert? 4

Habitats in an arid place 5

How deserts are formed 6

How deserts are changed 7

Different sorts of deserts 8

Where deserts are 9

ARID-ZONE PLANTS

Australia's arid-zone plants 10

Growing in a desert 11

The cycle of plant life 12

Remarkable desert plants 13

ARID-ZONE ANIMALS

Australia's arid-zone animals 14

Adapting to deserts 15

Desert survival strategies 16

Breeding after rain 17

TYPES OF AUSTRALIAN DESERTS

Tall eucalypt shrublands: the mallee 18

Tall acacia shrublands: the mulga 20

Hummock grasslands: spinifex country 22

Ranges and gorges: ranges, breakaways and jump-ups 24

Tussock grasslands: the Outback plains 26

Sandy deserts: sand dunes and sandplains 28

Shrub steppes: bluebush and saltbush plains 30

Ephemeral wetlands: salt lakes and claypans 32

Stony deserts: gibber plains 34

Desert coasts and islands 36

OUR FASCINATING DESERTS

Water in the desert 38

Plagues, poverty, plenty 40

Ferals in the desert 42

People and deserts 44

GLOSSARY (indicated in text by [G]) 46

INDEX TO SUBJECTS ILLUSTRATED 47

SOME INFORMATION SOURCES 48

The voyage of Ark Australia

About 160 million years ago, the world continent called Pangea split into two supercontinents, Laurasia and Gondwana.

By 70 million years ago, India, New Zealand and South America had broken away from Gondwana, leaving Antarctica and Australia behind. By 40 million years ago, Australia was drifting northwards on its own. Like a giant Noah's Ark, the new continent carried a cargo of living things. It was much wetter than today.

Over long periods of time, the continent grew warmer and drier. Wetlands disappeared. Rainforests shrank to small stands. From around 30 million years ago, drought-resistant plants became the most common floraG. Many animals adaptedG to survive in dry conditions. The ones that could not adapt died out.

Then, perhaps as much as 60 000 years ago, the ancestors of today's Aboriginal Australians crossed the narrowing sea from Asia and rapidly spread over the continent. They altered the balance of animal and plant life by their hunting methods and use of fire. Again, plants and animals adapted to this "firestick farming" (see p. 13).

The endangeredG Bridled Nailtail Wallaby is one of a group of small and medium-sized mammalsG that were once common in Australia's deserts and semi-arid lands.

In 1788, European humans came on board the Ark. With axe and plough they cleared large parts of Australia. They also introduced cats and foxes, rabbits, sheep, cattle and pigs. In just over 200 years great changes took place. The desert areas in particular suffered from overgrazing and attempts to farm their poor soils. NativeG plants and animals were lost, soils blew away and salt levels in the soil rose. The flow of inland rivers was changed by dams.

Gradually, Australians have come to realise that the Ark's resources need to be saved. They hope that some of the damage to its systems can be repaired and that some areas can be preserved as wilderness.

At present, Ark Australia still has on board much of its cargo of uniqueG living things. However, some of them are gone forever, and some survive only in tiny, protected areas. Some are still being destroyed.

This series of books describes the places where plants and animals live in Australia today – the habitats of Ark Australia. This volume is on deserts – the arid and semi-arid lands of the continent.

After rare heavy rain, plants grow and flower quickly in the deserts of Australia's Red Centre. In the distance is Uluṟu.

What is a desert?

Desert is land that gets less than 350 mm of rain per year. Just over 70% of Australia is desert. Some deserts are very large, while others are quite small. Deserts can look very different. Some have sand dunes, some are covered with spiky grasses, some are covered with scattered shrubs, or with rocks and stones, or with salt. The ground will have the colour of the rocks it is made from, and may be red, white, yellow, grey or brown. In one desert area, there may be several different desert types, forming a desert complex.

AUSTRALIA'S DESERTS CONSIST OF SEMI-ARID AND ARID LAND

SEMI-ARID LAND
250–350 mm of rainfall per year

ARID LAND
less than 250 mm of rainfall per year

Semi-arid land receives 250–350 mm of rainfall per year. In southern Australia, this rainfall may come in winter. In northern Australia, it comes in summer, from the edges of the wet season's tropical cyclones. Australia's semi-arid areas are often used for grazing stock.

Arid land receives less than 250 mm of rainfall per year. Sometimes the rainfall is far less than that amount. This rainfall may come at any time in the year. It often falls on small local areas in the form of brief, heavy showers. The picture above shows sand dunes and, at top right, a claypan[G].

FACTS 'N' FIGURES FILE

• **About 20%** of Earth's surface is desert. Human activities expand this every day.

• **Australia is the second** driest continent in the world. The driest continent is Antarctica, which is a cold desert.

• **About 35.5% of Australia** is arid land (less than 250 mm rain per year). It is home to only 0.1% of Australia's people.

• **About 37% of Australia** is semi-arid land (250–350 mm of rain per year). It is home to 3% of Australians.

• **Australia's Simpson, Great Sandy** and Great Victoria Deserts have sand dunes up to 300 km long and 300–400 m apart.

• **Highest temperature** so far measured in Australia's desert lands was 53.1°C at Cloncurry, Q, on 16 January 1889.

• **Low moisture** in the air means a desert gets twice the amount of solar radiation[G] as does an area with damper air.

• **A desert cools** quickly at night because there is no moisture to hold heat close to the Earth. Frost is common on winter nights in Australian deserts.

Habitats in an arid place

The Three-lined Knob-tail Gecko shelters in spinifex.

A cliff rises above boulders, spindly eucalypts and mounds of spinifex, a grass with leaves like needles. This arid place is home to many creatures. Its biodiversity increases greatly if water is available in a rock hole or spring.

The Common Wallaroo lives in rocky places. It shelters in caves.

Termite mounds house termites and other creatures.

The Thorny Devil eats small ants it finds on bare ground.

Ants make their nests underground or in cool, dark places.

The Woma, a large python[G], lives in burrows and caves.

Bilbies and many other animals shelter in burrows during the day.

BIOLOGICAL BUZZ-WORDS

- **Bios** was the Greek word for life.
- **Biodiversity** is the variety of different sorts of organisms that live in a particular place. Deserts have low biodiversity compared to rainforests.
- **Biomass** is the weight of living things in a given area. Deserts have low biomass compared to wetlands.
- **Biologists** study living things.
- **Biota** are living things. They are also called living organisms.
- **Botanists** are people who study plants.

- **An environment** is the total surroundings in which a living thing exists.
- **Ecologists** are people who study living things in their environments.
- **An ecosystem** is a community[G] of organisms.
- **The habitat** of an organism is the part of an environment in which it can survive most comfortably.
- **A niche** is a special place within a habitat that suits just a few species (it may also be called an **ecological niche**).
- **Zoologists** study animals.

How deserts are formed

Deserts form where there is little rain and the water that does fall evaporates[G] quickly.

Rain is a part of a cycle[G] by which water moves from the Earth's oceans into the air, onto the land, then back to the ocean.

The air holds water vapour[G] as tiny droplets that form clouds. Warm air can take up and hold far more water than cold air can.

When warm air that holds moisture is forced to rise, it becomes cooler. The droplets of water clump together to form raindrops that then fall to the ground.

In a warm desert, such as there are in Australia, the dry daytime air quickly evaporates any free water from the land. Then winds blow the moist air away before it can rise, cool and drop the water as rain.

A cold desert, such as the one that covers Antarctica, has its moisture locked up in snow and ice.

WHY DESERT WINDS BRING LITTLE RAIN

Winds reaching a desert may bring little rain because:

- Many deserts are in rain shadows. As air rises over mountains, it cools and loses its water vapour as rain. As the air flows down the other side of the mountains, it becomes warmer and sucks up moisture, making a rain shadow.

- Winds blowing over a cold ocean current are cooled. They pick up little moisture and carry little rain onto the land.

- Winds reaching the interior of a large landmass have already dropped most of their rain. They have little left to drop on the continent's centre, which becomes desert.

Clouds from the sea rise over a mountain range. Rain falls on the seaward slopes, while the landward slopes are much drier.

WHERE WORLD DESERTS ARE

Camels still provide transport in the sand deserts of the United Arab Emirates.

The world's great warm deserts are found in two bands around the globe between latitudes 10° and 30° north and south of the Equator.

The southern band includes the Atacama Desert (Peru), Namib and Kalahari Deserts (southern Africa) and the Australian deserts.

The northern band includes the Gobi Desert (China), the Thar Desert (India), the Sahara Desert (northern Africa), the Arabian deserts and the Mojave and Sonora Deserts (North America).

How deserts are changed

Blown sand Wind carries grains of sand that are blasted against rock, wearing it away. Dust and sand are carried away on the wind. They fall to Earth when the wind dies down.

Temperature changes Day's heat and night's cold cause minerals[G] in rock to swell and shrink at different rates. The surface of the rock may flake off, or the rock may split apart.

Action of water Rain carves the desert soil into gullies. Run-off carries mud and stones with it. As its flow slows down, it drops its load. Salt remains after salty water dries up.

Actions of animals Humans, domestic stock and feral[G] animals destroy plants and cut up the surface of the desert. This opens it to erosion[G].

A DESERT LANDSCAPE

Left: An eagle's view of a central Australian landscape. The trees growing in the bed and on the bank of the dry river at left and top suck up underground water with long roots. In the foreground is a mesa whose hard, flat cap has resisted weathering, while the softer rock below was cut away. Ridges of rock and sand, scattered with bushes and spinifex, fill the loop formed by the river. There is a water-carved gorge, or wadi (*close-up at right*), running across the neck of the river loop.

Different sorts of deserts

Tall eucalypt shrublands The Mallee.

Tall acacia shrublands The Mulga.

Hummock grasslands Spinifex country.

Ranges and gorges Ranges, breakaways and jump-ups.

Tussock grasslands
The Outback plains.

Sandy deserts
Sand dunes and sandplains.

Shrub steppesG
Bluebush and saltbush plains.

EphemeralG **wetlands**
Salt lakes and claypans.

Stony deserts The gibberG plains.

Coastal deserts Desert coasts and islands.

8

Where deserts are

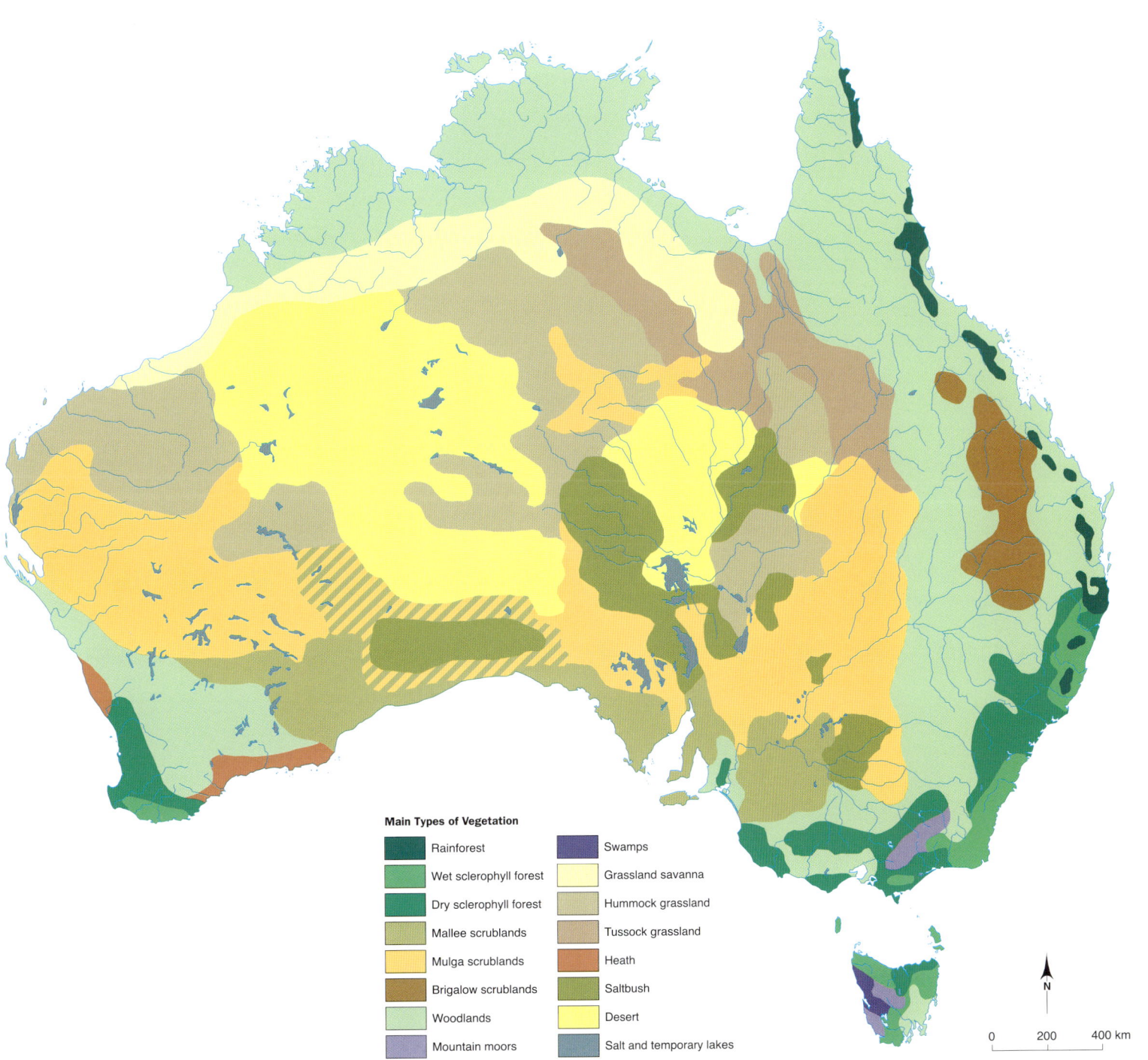

Main Types of Vegetation

- Rainforest
- Wet sclerophyll forest
- Dry sclerophyll forest
- Mallee scrublands
- Mulga scrublands
- Brigalow scrublands
- Woodlands
- Mountain moors
- Swamps
- Grassland savanna
- Hummock grassland
- Tussock grassland
- Heath
- Saltbush
- Desert
- Salt and temporary lakes

N

0   200   400 km

Australia's arid-zone plants

The cells of plants contain a green substance called chlorophyll. This allows the plant to make use of sunshine to power food-making. The raw materials the plant needs to make food include water and the elements[G] carbon (taken from the carbon dioxide that exists as a gas in the air) and nitrogen (dissolved in water, taken from the soil into the plant). For healthy growth, plants also need some calcium and phosphorus, as well as tiny amounts of copper, magnesium, zinc and other trace elements. Australian soils carry low levels of these plant nutrients[G].

HERBS, SHRUBS AND TREES

This desert area contains herbs, shrubs and trees.
Herbs are plants with non-woody stems.
Shrubs are multi-stemmed or single-stemmed woody plants with branches close to ground level.
Trees are single-stemmed woody plants over 5 m tall when grown. The stem of some trees separates into two or more trunks just above the ground.

LIFE'S A BATTLE

The tree shown above is growing on a rock face. It gets water and nutrients through its roots that creep into every crack in the rock (left).

TOUGH-LEAVED PLANTS

Many of Australia's desert plants have tough, stiff leaves spaced a long way apart on their stems. They may grow that way because of low levels of nutrients in Australian soils.

These tough-leaved, or sclerophyllous, plants began to develop in Australia at least 50 million years ago. They include eucalypts, banksias, grevilleas, she-oaks, paperbarks and acacias.

Desert plants get nutrients from soil enriched by rotting plant or animal material. Nutrients may also be washed into the area by heavy rain, or carried there by wind.

This eucalypt grows between sand dunes. Leaves and bark that it has shed enrich the soil beneath it.

Growing in a desert

DESERT PLANT SURVIVAL TECHNIQUES

A desert tree may grow very long roots that suck up water from deep under the ground.

Rain drips from the leaves of a spreading plant and is sucked up by a mat of fine roots under the plant.

Parasitic^G plants, such as mistletoes, sink feeding roots into other plants and use their water and nutrients.

Fine hairs on leaves and stems trap moisture and hold it to form a micro-climate^G around the plant.

Fleshy leaves and stems store water. In a drought, the plant can use this stored water for survival.

Some plants, such as spinifex, have leaves that are hard, spiky needles that conserve water.

FACTS 'N' FIGURES FILE

- **Two hundred thousand** seeds were found in 1 m² of desert soil.
- **"Resurrection plants"** dry up and appear dead in times of drought. After rain falls, they spring to life.
- **Plants lose water** through tiny pores on their leaves and stems.
- **Xerophytes** are plants that have changed their structure to help them survive extreme heat and lack of water.

The silvery sheen on the leaves of some desert plants reflects the sunlight so the plant stays cool.

Some desert trees have drooping leaves. These shade each other and do not offer much area to the Sun.

The cycle of plant life

Desert plants need to be able to reproduce[G] themselves in the small window of opportunity created by rainfall.

Some desert plants live for a long time, waiting out dry years as dead-looking, leafless skeletons, or as underground tubers[G] or bulbs. Other desert plants are short-lived ephemerals that take only weeks to complete their whole life cycle.

Whether the plant is long- or short-lived, its seeds may lie on the ground, or in the soil, or be blown around by the wind, for two or three years before enough rain falls to make them sprout.

EPHEMERAL PLANTS – GROWN AND GONE IN A SHORT TIME

A sand dune before rain.

Part of the same dune some weeks after good rain.

WAYS OF SPREADING POLLEN AND SEEDS

A seed grows from a plant ovum[G] that has been fertilised[G] by a pollen[G] grain. Pollen may be carried from one flower to another by the wind, by water or by an animal.

In order to grow well, a seed needs to be carried away from its parent plant. The seed may travel on the wind or in water. It may be carried on the coat of an animal. Some seeds are packaged in tasty fruit, so they are spread in animal droppings. Some spring open and jump away from their parent plants.

The pollen of grasses is carried in the wind or by insects.

To get nectar[G] from this flower, a bird must brush past pollen-bearing stamens.

Quondong fruits are tasty. The flesh is eaten, the seed is dropped.

Remarkable desert plants

WAVY MARSHWORT

This grows in still or slow-moving water with a muddy bottom. Its leaves spread in a floating mat. It grows quickly when claypans in the desert fill after rain. When the claypan dries up, the small, shiny, straw-coloured seeds remain. They will grow when the claypan fills with water again.

HONEY GREVILLEA

This belongs to a group of plants that includes the banksias and hakeas. It has thin, tough leaves that conserve water. After rain, its flowers appear in spikes. They contain nectar that is used by the Aborigines to make a sweet drink. Birds and other animals feed on the nectar and carry the pollen to other flowers.

GREEN BIRDFLOWER

A member of the pea family, this plant is sometimes called Rattlepod because of the noise its seeds make. Like all peas and beans it has bacteria[G] on its roots that help it get nitrogen from the soil. If horses eat this plant, they develop a sickness called "walkabout", which can kill them.

RED CABBAGE PALM

Found in Palm Valley in central Australia, this tree grows nowhere else in the world. The nearest relative is 1000 km away. These palms have remained in this sheltered valley from a time when central Australia was much wetter than it is today. They get their water from springs flowing from the sandstone of the gorge.

FIRE – FRIEND AND FOE

Many Australian plants can grow again after a bushfire sweeps over them. They are protected by thick bark, or they sprout again from buds on their stems, or they grow again from special underground stems. Some Australian plants need heat or smoke to open the hard capsules[G] that protect their seeds.

The Aboriginal people used "firestick farming" to burn off old growth and bring on green shoots. These attracted animals that could be hunted.

Australia's arid-zone animals

All animals need air, water and food. They need ways of sheltering from too much heat and too much cold, and ways of protecting themselves against predators^G.

Like plants, desert animals have adapted their bodies so they are better able to survive in a dry, very hot or very cold place. They have also developed ways of behaving that protect them from too much heat, dryness or cold.

These adaptations interact. For example, many desert animals are active in the cooler night (a behavioural adaptation). Their eyes are large and give good night vision and their senses of smell and touch are keen (physical adaptations).

○ NIGHT-TIME HUNTER ○

During day, the pupil of a night-active animal's eye shrinks to a slit.

At night, the pupil of the eye opens wide to let in as much light as possible.

HIP-HOP

Desert animals need to save energy. Hopping takes less energy than running. When a kangaroo hops, its bulky gut flops up and down. This pushes air out of its lungs, then lets it rush back in, saving the energy that other mammals use in breathing.

A DESERT FOOD-CHAIN

All animals depend on plants for food. Herbivores eat plants, while carnivores eat herbivores. Scavengers eat dead animals. In Australia's deserts, the main herbivores are termites (small insects). The main carnivores are reptiles^G. The body wastes and carcasses of all animals decay and provide plants with nutrients to complete the cycle.

Termites.

Skink.

Legless lizard.

Spinifex.

Wedge-tailed Eagle.

Dingo.

FACTS 'N' FIGURES FILE

- **Desert animals** are often pale in colour, so they take in less heat from their surroundings.

- **Some animals lose heat** through parts of their bodies, such as large ears that contain many blood vessels.

- **Panting and licking** parts of the body help some animals get cool.

- **About 250 species of ants** live in Australia's deserts.

- **Australia's deserts** are rich in reptile species. Arid Australia has 150 species of lizards. Only 57 species of lizards live in the deserts of North America.

- **Medium-sized marsupials**^G such as hare-wallabies and bandicoots cannot survive habitat destruction. Six desert species are extinct^G. Four others survive on islands.

- **A few species of endangered** desert mammals are being bred in captivity.

- **Some captive-bred desert mammals** may be released in their traditional habitats if cats, foxes and grazing by sheep and cattle can be controlled there.

Adapting to deserts

In summer, desert animals need to control the amount of water they lose from their bodies. They also need to prevent their inside body temperatures becoming too high. In winter, they need to keep their inside body temperatures warm enough for survival on cold, possibly frosty, nights. Desert creatures hide from both heat and cold in burrows or under plants or rocks. In cold weather they bask in the sun, preferably on dark, heat-holding surfaces.

In very hot or very cold weather, some desert animals go into a sleeping state. Their body functions slow right down. When the temperature is right, or when rain arrives, they wake up.

Reptiles like this Sand Monitor shelter in burrows and holes. In cold weather, they warm up by basking in the sun.

Kangaroos and wallabies shelter in caves and under bushes. They sun themselves on cold winter mornings.

Desert birds are most active in early morning and late afternoon. They rest in shade during the day's heat.

Southern Hairy-nosed Wombats hide from heat and cold in burrows. They come out at night to feed.

The Three-lined Knob-tail Gecko hunts at night. It stays active while the lizards it eats are slowed down by night's cold.

The Bridled Nailtail Wallaby spends the day in a hollow scooped under a spinifex hummock.

Desert survival strategies

There are three main ways for desert animals to survive their region's hot, dry summer and cold, dry winter.

- They can adapt their bodies to deal with lack of water and food and extremes of temperature.
- They can behave in ways that give them the best chance of survival.
- They can go somewhere else where conditions are better, returning when rain falls.

Small creatures usually adopt the first two strategies. Large mammals and birds can add the third strategy.

SLEEPING BEAUTIES

Some desert creatures survive dry, hot or cold times by having sleeping, or dormant, stages in their life cycles.

- Desert insects lay eggs that remain dormant in dry and cold periods. They hatch when rain falls.
- Adult frogs become dormant in dry weather, sleeping in a burrow or under a rock.
- Reptiles are active in spring and summer, but may be dormant during winter.
- Some small mammals that feed on insects become torpid^G when cold, windy weather prevents them hunting. Short-beaked Echidnas may also shut down body activity in very cold weather.

During the day, the mouse-sized Fat-tailed Dunnart shelters in a burrow, coming out to feed in the cooler night. On cold nights when insects are scarce, the dunnart may become torpid for up to 12 hours.

A number of desert frogs survive dry spells in a burrow. Protected by a coat of water-conserving mucus^G, they use water stored in their bladders. When rain falls, they emerge.

THE WANDERING LIFE

The White-browed Woodswallow is one of many nomadic birds that stay in an area while food is plentiful. When the country dries up and insects become scarce, these nomads move on.

WATER-SAVING BODIES

Spinifex Hopping-mice get little water from the dry seeds they eat. They breathe out little moisture and their bodies recycle most of the water from their urine and droppings.

Breeding after rain

Australia's desert animals do not get the opportunity to breed very often.

Many have to wait for rain to make food plants grow and will breed after rain falls, no matter what the time of year. When rain falls in an area, the animals that live there come quickly into breeding condition.

Some desert animals are long-lived because they may only get the opportunity to breed a few times during their lifetimes. Others have short lives and so do not compete for food with their offspring.

Female desert animals may have small numbers of babies compared to females of the same species in wetter regions. They may stop having babies altogether in drought times.

Birds may fly long distances – for example, from the coast to central Australia's salt lakes – to breed. They fly back again after raising their chicks.

FAMILY AFFAIRS

This male Splendid Fairy-wren is part of a group feeding chicks.

The desert is a tough environment that encourages cooperative behaviour. Every member of a group contributes to the group's well-being. One pair of birds cannot gather enough food to bring up a brood. Because of this, the young of many desert birds stay with their parents and help them feed further broods of chicks.

○ LONG FLIGHTS ○

Some waterbirds leave the coast to breed on newly filled desert lakes. They feed on the shrimps and fishes that are there in vast numbers.

The Australian Pelican flies long distances to nest on desert lakes.

○ SHORT LIVES ○

Some desert lizards live for less than a year and are all dead by the time their eggs hatch. This means they do not compete with their babies for food.

Eggs laid by some desert lizards may hatch after their parents die.

Red Kangaroos.

A PAUSE IN PREGNANCY

A female kangaroo can mate again one day after a joey is born. The new embryo⁶ will grow to about 0.25 mm across, then stop developing. It will start developing again when the joey leaves the pouch or if it stops suckling. Kangaroos keep breeding while food and water are available. In times of drought, few young survive.

FACTS 'N' FIGURES FILE

• **The eggs of the plague locust,** a type of grasshopper, must take in twice their weight in water before they can hatch.

• **The eggs of the tadpole shrimp** can survive for years in the ground. When rain floods the claypans, they hatch. The shrimps lay new eggs after 7 days' growth.

• **Australia has** over 60 species of birds that have cooperative breeding behaviour.

• **Aboriginal people living in deserts** dug up water-holding frogs and drank the water stored in their bodies.

• **Harvester ants** collect and store seeds in good times. They live off their stores until it rains again. In some places in Australian deserts there is a nest of these ants every 1–2 m².

• **A number of lizard species** may share a desert habitat without competing for food. They eat different sorts of food, hunt in different places and are active at different times of day and night.

• **In drought times,** more than half of the Red Kangaroo pouch joeys in an area will die before they reach 3 months of age.

• **After rain falls,** native rats and mice breed in huge numbers in arid areas.

Tall eucalypt shrublands

THE MALLEE

Mallees are eucalypts that grow from a swollen root called a lignotuber.

Above the ground the trunk divides into a number of branches that carry a flattened canopy[G] of leaves. In the wetter semi-arid areas, the crowns of mallee trees may almost touch. Where it is drier, they are farther apart, with low shrubs, grasses and spinifex growing between them.

Mallee grows most often on sandy soils and may be found on the ridges of sand dunes.

Mallee country is used for grazing stock and is cleared to grow cereal crops such as wheat. However, the rainfall is often unreliable and clearing mallee country may lead to soil erosion.

The dominant[G] trees in mallee country are eucalypts.

A lot of the Mallee has been cleared to grow wheat.

BULL AND WHIPSTICK MALLEES

The number of branches on a mallee tree depends on the soil type and how often fire burns across the area.

Bull mallees, with just a few quite strong branches, grow on good soil and/or where there are few bushfires.

Whipstick mallees, with many thin, bending branches, grow on poor soil and/or where there are many bushfires.

FACTS 'N' FIGURES FILE

- **Clearing mallee country** was difficult for early farmers. Plough blades broke on the tough lignotubers. Burning them made them send out shoots. Mallee soil was poor in plant nutrients.

- **In the 1890s,** the invention of the stump-jump plough and the use of super-phosphate fertiliser[G] allowed large areas of mallee to be cleared for wheat-growing.

- **The Mallee is the name** given to an area of mallee country in western Victoria.

- **A large area** of the Victorian Mallee was cleared for wheat-farming. In the 1979–83 drought, huge clouds of dust blanketed Melbourne as the surface soil of the Mallee blew eastwards.

- **About 12 species of birds** live only in mallee. Most could live nowhere else.

- **The female Mallee Fowl** spends a great deal of energy producing a large number of big eggs. She may help the male open the mound but otherwise does no work on it.

- **If an adult Mallee Fowl** scratches up a chick as it struggles to the surface of the mound, it takes no notice of the baby.

- **Mallee charcoal** is still in demand for barbecues and the charcoal chicken industry.

- **After mallee is cleared,** groundwater brings salt to the surface.

- **Restoring salted mallee land** involves planting trees and salt-tolerant shrubs that will help lower the water table again.

Tall eucalypt shrublands

FOUND IN AND AROUND MALLEE

Galahs.

Jewel beetle.

Mallee Ringneck.

Singing Honeyeater.

Grey Kangaroo.

Short-beaked Echidna.

PROTECTING HABITAT

The Little Desert covers an area of mallee 25 km by 100 km in Victoria's central west. Its poor soil saved it from clearing for wheat farms.

In 1955, about 200 ha were put aside to save 14 Mallee Fowl mounds. In 1968 the area was enlarged to 945 ha and named the Little Desert National Park.

A government plan to clear 80 000 surrounding hectares for farming was opposed by local people. In 1969 the park was increased to 35 300 ha. The National Park preserves plants and animals that are gone from the surrounding farms.

EGGS IN A NATURAL INCUBATOR

The Mallee Fowl is a large, ground-living bird found in widely separated areas of mallee across southern Australia. Clearing of the mallee, sheep and rabbits grazing the plants it eats and predation[G] by foxes have made it rare.

A mated pair stays together for life.

In a sandy clearing, the male scratches soil and ground litter into a mound. He then rakes leaves into a central crater[G] and covers them with soil. He opens the nest so the plant matter is dampened by rain, then scratches sand over it. As the vegetation rots, it makes heat.

The female lays 3–33 eggs at around 6-day intervals into the nest-chamber. After each one is laid, the male covers it with sand. He then opens the nest, or rakes on more sand to keep the nest chamber at around 34°C. The chicks hatch after about 63 days, burrow out of the mound and run away. They can fly within 24 hours and can feed themselves, and have nothing to do with their parents.

A male Mallee Fowl opening the nest chamber in his mound.

Mallee Fowl egg-eaters (left to right): Shingleback, Sand Monitor, fox.

Tall acacia shrublands

THE MULGA

Mulga is a species of acacia (or wattle) that grows in many desert places. Mulga country may also contain many other sorts of acacias beside mulga.

Some acacias have small leaflets. Others have adapted to dry conditions by flattening their leaf stems and doing away with leaves. The leaf stems, which are known as phyllodes, transpire[G] less than true leaves would.

Rain is channelled along the phyllodes to the branches, then down the tree trunk onto the root zone. A layer of roots near the surface collects this water. Another layer of roots further down soaks up water held in the lower levels of the soil.

Mulga country.

An aerial view of Mulga country.

Acacia phyllodes and flowers.

A Ground Cuckoo-shrike nesting in the twisted branches of an old Mulga tree.

Mulga grows from seeds that develop in pods. Three separate falls of good rain are needed to make Mulga flower, then set seed, then sprout the seeds. These conditions occur at around 10-year intervals. The seedlings are readily eaten by cattle, sheep and rabbits.

A Mulga tree flowers after rain has fallen.

FACTS 'N' FIGURES FILE

- **Acacias have only grown** in Australia for about 30 million years. Seeds probably drifted to Australia on sea currents.

- **About 835 species** of acacia occur in Australia.

- **Bacteria** that live on the roots of acacias take nitrogen from the soil. This plant nutrient can be used by the acacia and by other plants nearby.

- **Acacia seeds** have fleshy, often brightly coloured and tasty stalks. Ants store them in their nests where they are surrounded by moist, nutrient-rich soil.

- **The Aboriginal people** roasted and ate acacia seeds. The bark of acacias was thrown into pools to stun fish.

- **One of the first Australian plants** known to Europeans was an acacia. It was probably collected[G] by Willem de Vlamingh near the Swan River, WA, in 1697.

- **Where more than 300 mm** of rain falls in a year, on average, Mulga is replaced by eucalypts.

- **A Mulga tree can survive** all but the worst droughts and may live 250 years. However, most acacias live less than 20 years.

Tall acacia shrublands

THE MULGA BLOSSOMS

After rain falls on Mulga country, the ground becomes covered with flowering plants. Birds flock to the area to eat flower nectar, fruits, seeds and insects. They breed while the good times last.

Poverty bushes (also called emu bushes) blossom after rain.

A Crimson Chat flies to his nest in a mulla mulla bush.

A minirichie tree that has survived many dry years is surrounded by ephemeral Everlasting Daisies.

There are many sorts of desert peas.

WITCHETTY BUSHES

An Aboriginal woman digging for witchetty grubs.

Witchetty grubs (above) are the larvae^G of a large moth. They bore into the roots of an acacia known as the witchetty bush. They are used for food by desert Aboriginal people.

USEFUL ACACIAS

A Bearded Dragon uses a mulga-wood fence post as a perch.

Acacia trees are very useful to humans. They provide seeds full of protein^G and tough wood for spears, dishes, buildings and fence posts. Their leaves give shade and can be eaten by sheep and cattle. Their bark can be used to tan leather.

PLAYBOY BIRD

The male Spotted Bowerbird makes a stick avenue under a drooping acacia. He sings and dances there to attract females.

A male Spotted Bowerbird arranges bones outside his bower.

Hummock grasslands

SPINIFEX COUNTRY

Grasses are plants whose leaves, also called blades, have veins lying side by side instead of forming a network. Their seeds have one seed leaf, rather than the two found in plants such as peas, eucalypts and acacias.

Creeping grasses put out long, ground-hugging stems. These form roots and leaves from joints called nodes. They are found growing in sand and are used as pasture and lawn grasses.

Tussock grasses form clumps or hummocks. Their leaves grow from their bases, not their tips.

This means that the tussock grass leaves can be eaten, mowed or burned and will then grow again.

There are around 700 species of native Australian grasses. Some grasses are found in nearly all Australian plant communities. The spinifexes are tussock grasses that have adapted to desert conditions and have tough, narrow leaves. The spiky leaved species are called hard spinifexes. The limper leaved species are called soft spinifexes. Between 15 and 20% of Australia is covered with spinifexes and other arid country herbs.

HARD SPINIFEX

Hard spinifex forms half-circles on this stony hillside.

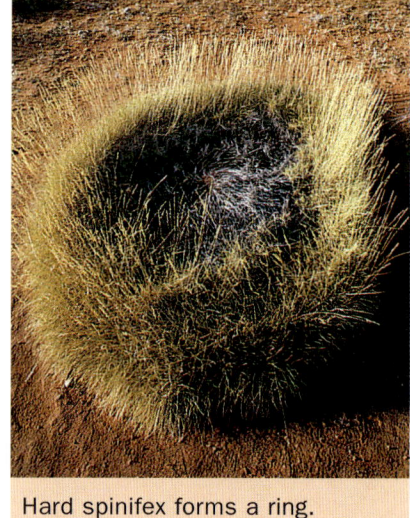

Hard spinifex forms a ring.

Spinifex flowers and seeds are on tall stalks.

An aerial view of spinifex.

Hard spinifex is a spiky grass that forms a low hummock. The leaves are sharply pointed. New growth rises on the outside of the hummock, so when the old growth dies or the hummock is burned it becomes a ring. This may be up to 10 m in diameter. Spinifexes burn very easily. Many native animals live in spinifex or burrow under it. They are protected by its spiky leaves.

Hummock grasslands

ANIMALS IN THE SPINIFEX

The Centralian Blue-Tongued Lizard eats plants and small creatures. It is active during the day.

The Spinifex Pigeon eats the seeds of spinifex and other plants. It needs to drink each day and lives near water.

The Yellow-faced Whipsnake feeds mainly on small lizards. It hunts during the day and is very fast-moving.

The Spinifex Hopping-mouse lives in a burrow with other hopping-mice. At night they all come out to eat seeds.

Many creatures live in spinifex country. The hummocks shelter their homes, the spiky leaves protect them from predators and there are plenty of seeds from spinifex and other plants to eat. The herbivores range in size from termites and other insects to wallabies and kangaroos. The main predators are lizards and snakes.

THE VERY RARE ONES

Rufous Hare-wallaby (also called a Mala).

Bridled Nailtail Wallaby.

The Rufous Hare-wallaby and the Bridled Nailtail Wallaby once lived in spinifex country. They were hunted by humans and foxes, and the grasses they ate were grazed out by sheep and rabbits. Today they are almost extinct. Efforts are being made to breed them in big, open-range zoos in arid country. If the places they once lived in can be protected from predators and from animals that compete for food, these creatures could be seen again in the wild.

23

Ranges and gorges

RANGES, BREAKAWAYS AND JUMP-UPS

Australia's desert ranges are not high. They were formed more than 300 million years ago and have been worn down into stumps. However, they provide habitat for an amazing variety of animals.

One reason is that water gathers in pools in steep-sided gorges. Rain may top up waterholes where rivers once ran. Springs may be fed by water trickling through cracks between layers of rock, or soaking through porous[G] rock. Caves in cliffs and under overhanging rocks provide shelter for many creatures. In the cool, late afternoon they come out to feed.

Some desert areas have a covering of rock hardened with silica[G] or iron. This is called a duricrust. The surrounding country may weather away, leaving a large flat area called a mesa, or a smaller one called a butte. The edges of these are known as breakaways or jump-ups.

The MacDonnell Ranges stretch for 400 km. They are the worn-down remains of mountains that were once as high as the Himalayas.

Rivers have cut gorges into Australia's desert ranges. The gorges form habitat for many plants and animals.

A breakaway is the edge of a tableland. It may also be called a jump-up.

This pillar (butte) is capped with duricrete.

A water-loving rush plant grows where a spring makes the rock damp.

Ranges and gorges

ROOS ON THE ROCKS

A Yellow-footed Rock-wallaby.

A Wedge-tailed Eagle attacking a Common Wallaroo.

Rock-wallabies and wallaroos live on rocky hillsides. The Wedge-tailed Eagle, which soars[G] on warm air rising from the ranges, may prey on these animals and their young ones.

GHOST STORY

Ghost Bats live in desert caves and mine shafts. They eat small animals, including other bats. They are rare and will leave their daytime roost forever if disturbed by humans.

RANGE REPTILES

The Ridge-Tailed Monitor lives on breakaways and among rock outcrops. It eats insects and lizards.

At night, the Marbled Velvet Gecko leaves its hiding place between rocks. It hunts insects across the open rock faces.

FACTS 'N' FIGURES FILE

• **Some rocks in central Australia** were formed nearly 2000 million years ago. The Pilbara of Western Australia contains rocks around 3500 million years old.

• **Between 1870 and 350 million years ago,** central Australia was covered by a shallow sea with sandy beaches. Sand washed into the sea and hardened to form sandstone. Some of this was changed by heat into a rock called quartzite.

• **The MacDonnell Ranges** were pushed up between 340 and 310 million years ago. They were once over 9 km high.

• **Desert ranges** that may rise to 1000 m from the plain are home to many plant species. West MacDonnell National Park has almost 600 different sorts of plants.

• **Gorges in desert ranges** are homes to birds that need to drink every day, such as pigeons, finches and parrots.

• **Ranges are refuges** for threatened species such as rock-wallabies. However, in some places the wallabies are battling for food against feral goats. Not only that, the goats take over the rock shelters that the wallabies use as protection from predators such as eagles.

Tussock grasslands

THE OUTBACK PLAINS

Most of Australia's native grasses grow in tussocks. They include wallaby grass, kerosene grass, spear grass, Mitchell grass and the spinifexes. In the north of the country, tussock grasses may grow over 1 m in height. Other plants will grow between the tussocks.

Grasslands form one of Australia's most threatened ecosystems. Sheep and cattle do well on native grasses in good seasons. In drought, and even in normal years, they overgraze the pasture and trample areas near water. Rabbits and feral pigs add to the damage. This has led to the disappearance of some native animals that depended on long grasses for shelter and food.

Where trees are destroyed, salt comes to the surface, spoiling the ground for many plants.

An aerial view of tussock grassland shows gaps of bare soil between the tussocks of grass.

Australia's tussock grasslands are used for grazing stock.

FACTS 'N' FIGURES FILE

- **The Prairies (North America),** Pampas (South America), Steppes (Eastern Europe and Asia) and Savanna (Africa) are all grasslands. Most have rich soil.

- **Native grasses** need less water and recover more quickly after bushfires than do introduced grass species.

- **Most remaining grassland** in southern Australia is on private land. It must be bought before being protected.

- **A number of medium-sized marsupials** and one bird (the Night Parrot) once found in Australia's grasslands are now extinct.

- **The Double Gee** is a spiked burr brought to Australia's grasslands from South Africa.

- **When Aboriginal people burned the grasslands,** they were careful to fire only small areas that had not been burnt for some time.

- **Five million hectares** of the Mitchell-grass plains of Queensland are threatened by a prickly acacia introduced in 1890 as feed and shade for stock.

- **Termites** and their mounds are common on grasslands.

HERALDIC BEASTS

The two animals that hold up the shield in the Australian Coat of Arms are both found on the Outback plains. The Red Kangaroo (above right) and the Emu (right) are mobile^G grazers that appear in an area when rain makes the grass shoot. When the grass is eaten, they move to other areas.

Tussock grasslands

Major Mitchell Cockatoos eat seeds and grains.

The Australian Bustard eats insects and plants. It is a rare bird.

A Hooded Parrot perches above its nest hole in a grassland termite mound.

The Australian Kestrel is a small falcon that hunts grasshoppers, mice and other small creatures.

The Black-breasted Buzzard breaks open the eggs of ground-nesting birds and eats the contents.

Fat-tailed Dunnarts hunt small creatures around the grass tussocks.

The Grass Owl is a plains predator. It nests amongst grass tussocks.

Tiny carnivorous marsupials such as planigales can slip into cracks in the ground.

Sandy deserts

SAND DUNES AND SANDPLAINS

Sand is made up of fine grains of a mineral called silica. The red colour of Australia's desert sand comes from a fine coating of iron oxide. Sand is carried across a desert by wind or water. Larger grains skip or bounce along. As they land, they dislodge other grains that join them. This chain reaction can set an entire sand surface in motion. The stronger the wind, the more sand it can carry and set in motion.

When the wind slows down, the sand drops to the ground. It piles up to form dunes, or ridges, that gradually shift downwind. Once a small dune starts to grow, it acts as a barrier to the wind, so that sand drops on its slopes.

Australia's sand dunes range from small ones a few metres high to as much as 200 m high and 1 km wide. Between them are depressions, or corridors, where water collects after rain. Different plants grow on the crests of dunes, on their slopes and in the corridors between them.

These ripples are caused by wind action. Plants grow where most moisture gathers on a dune's surface.

The plants on this dune bend downwind in the same direction that sand ripples form.

Between sand dunes there may be quite a thick growth of plants. The Dingo is a predator of sandy deserts.

Animals that live in sandy deserts often feed at night. They leave their tracks to tell their tales.

Sandy deserts

ANIMALS AND PLANTS OF THE SANDY DESERT

The rare Bilby lives in a burrow. It eats plants and seeds at night.

The seeds of sand-dune plants may be carried by the wind.

The Burrowing Bettong once lived in the damper areas between desert sand dunes.

The blind Marsupial Mole lives underground. It eats insects.

Parakeelya leaves store water. The seeds were eaten by Aborigines.

The Thorny Devil drinks dew that collects on its skin at night. It eats ants.

This hunter wasp will lay eggs on its paralysed[G] spider prey.

Sturt's Desert Rose is the emblem of the Northern Territory.

Honeypot ants store food for their colony[G] in their swollen abdomens.

FACTS 'N' FIGURES FILE

- **A sand dune** complex in central Australia may be home to 40 species of lizard. Many of these species eat termites.
- **The Marsupial Mole** digs with two flattened claws. As it swims through the sand, the burrow fills in behind it.
- **The Burrowing Bettong** once lived in most of Australia's drier areas. Today it is found in the wild only on four small islands off the north-west of Western Australia.
- **The Desert Bandicoot** lived in sandy north-western deserts. The last one seen was collected in 1943.
- **The Bilby** makes a burrow that may be 3 m long and nearly 2 m deep. As a predator digs into the entrance, the Bilby tunnels away from the closed end.
- **Hunter wasps** dig burrows and stock them with live, paralysed spiders or insects. When the wasp larvae hatch, they feed on the stored prey.
- **Parakeelya flowers** open for only one day. A bush has a long taproot. It can live and flower on the moisture in its leaves.

29

Shrub steppes

BLUEBUSH AND SALTBUSH PLAINS

Bluebushes, saltbushes, samphires and bindiis belong to a group of shrubs called chenopods. This name means "goosefoots" (from the shape of the seed cases). They are low shrubs up to 1.5 m high that grow on land that may be stony, limey or salty. Often their leaves are fleshy, or their stems are jointed and fleshy and their leaves very small. Both leaves and stems may be covered with fine hairs, tiny salty scales or a floury substance. Some of these plants live for only one season, others may live for many years.

Chenopod shrublands grow on land that is quite rich in plant food, and they are heavily grazed by kangaroos, rabbits, sheep and cattle. In a normal season, the ground between the shrubs is bare. After good rain it becomes covered with ephemeral plants.

A chenopod shrubland. The silvery appearance of these bushes comes from water-conserving surfaces on their leaves.

The entrance to a cave on the Nullarbor Plain. The limestone plain is covered with chenopod shrubs and there are few trees.

Bindiis have spiky or spiny seeds that catch onto passing animals.

Ants live in nests beneath the hard soil capping of the plains.

The fleshy leaves of this saltbush hold water.

FACTS 'N' FIGURES FILE

- **There are many species** of bluebush. A grey, floury covering on their leaves makes them appear blue-grey.

- **The ash of the "glasswort"** chenopods was used to make sodium bicarbonate[G] for glass making.

- **Old Man Saltbush** grows to 3 m high. In arid country in southern Australia it is heavily grazed by sheep and cattle.

- **Bindiis** are chenopods whose seed-cases carry spikes, spines and hooks. These attach to animals that pass by.

- **Termites** are rarer in chenopod shrublands than in spinifex country. This is because spinifex is much richer in the woody cellulose[G] that termites feed on.

- **The Nullarbor Plain** is covered with chenopod shrubs. The "no tree" plain is surfaced with limestone and honeycombed with underground caves.

- **There are fewer lizards** in chenopod shrubland than in most arid areas.

- **The Aborigines** ate chenopod berries and ground their seeds into flour.

Shrub steppes

BORN TO WANDER

The Red Kangaroo roams the plains in search of fresh plant growth to eat. It can move long distances to find food.

The Black Falcon is a nomad that hunts smaller birds when they are plentiful, then moves on in harder times.

Nomadism helps some animals survive hard times in the desert. Many birds, the Dingo and the Red Kangaroo are able to leave an area when food is scarce. They move until they find a place where living is easier. They will return when conditions improve.

HOME-MAKERS MAY NOT SURVIVE

Greater Stick-nest Rats.

(Above) Southern Hairy-nosed Wombat. (Below) Burrow entrances.

Animals that cannot leave their usual home range[G] when poor conditions or competitors arrive have a smaller chance of survival than nomads.

The Greater Stick-nest Rat once built nests up to 1 m high and 1.5 m across on arid plains. It disappeared when sheep and cattle arrived and now exists only on a few offshore islands.

The Southern Hairy-nosed Wombat can survive heat and dryness in its moist burrow as long as it can find green food. Rabbits and sheep eat the native plants that baby wombats need for weaning from their mother's milk.

Ephemeral wetlands

SALT LAKES AND CLAYPANS

Salt lakes and claypans are places where water gathers, only to evaporate again in the desert heat. The water may come from local rainfall, or it may run into the lake or claypan from far away, through winding creeks and channels. Both salt lakes and claypans have beds of heavy clay soil that turns to mud when it is wet and stops too much water sinking through it. When it dries out, the mud cracks, then its surface turns to dust and blows away.

Some salt lakes have water in them most or all of the time. Claypans disappear after a time.

The salt in a salt lake comes from water that carries dissolved salt. This has been picked up from land where trees have been chopped down and salty groundwater has risen to the surface. When salty water dries up, the salt is left as crystals. Organisms that live around salt lakes must deal with the salt in their water and food.

The water in a claypan is usually fresh, but it is thick with tiny grains of mud or sand. When a claypan fills, it becomes a breeding place for frogs, fishes, shrimps, birds and insects, and a drinking place for mammals.

The surface of a drying lake.

A Red-necked Avocet nesting among samphire near a salt lake.

Salt lakes may be dotted with islands covered with samphire and other salt-tolerant plants.

Salt crusting desert sand.

Samphire grows on and around this small salt lake, also called a saltpan.

FACTS 'N' FIGURES FILE

- **The salt in Australia's soils** is there naturally. It has been said that 10 000 t of salt lie in soil and rock beneath each hectare of wheat-growing country.

- **Few plants grow** well in salty soil. If a farmer clears too many trees on the family property, the land can become too salted for the grandchildren to use.

- **A playa** is a salt lake.

- **A boinka** is a plain covered with salt and gypsum, a white mineral used to make plaster of Paris.

- **Close to a salt lake**, there are few plants. As distance from the lake increases, so does the number of plants and species.

- **Samphire** is a general name for a group of about 40 species of bushes with fleshy stemmed, leafless branches. Samphire may contain up to 24% salt.

- **When salt lakes fill** with water, tiny brine shrimps breed in huge numbers. Long-legged birds such as avocets and stilts fly in to feed on them.

Ephemeral wetlands

LIFE AND DEATH ON A CLAYPAN

A claypan will quickly fill after heavy rain.

This claypan has formed on clay soil between two sand dunes.

Above: Aquatic plants grow from seeds and roots. Waterbirds such as this Australasian Grebe fly in to nest on the claypan. Below, left to right: Animals drink at the claypan; if no more rain falls, the pan dries up; animals that cannot find water die.

Stony deserts

GIBBER PLAINS

The arid areas of Australia contain many mesas (flat-topped hills) and tablelands.

These are capped with hard rock called duricrust that was formed about 60 million years ago when the climate was much wetter. Minerals dissolved in water cemented rocks together and then formed new sorts of rock. Today's duricrusts may be cemented together by the minerals silica, iron oxide or calcium carbonate. Sometimes they contain lumps of iron and silica.

The gibbers, or rocks, of Australia's stony deserts are all that remains of completely worn away duricrusts. They are often stained red and purple by iron oxides.

Many of the minerals needed by plants are locked up in duricrusts. Plants cannot use them, and the tops of mesas and gibber plains often have little plant life. However, there are plants and animals that live in these hostile places; some of them are found nowhere else. Animals may live in burrows or caves beneath the duricrust, or where it makes an overhang at the edge of a breakaway. Animals that feed on gibber plains may be camouflaged[G] so they are difficult for predators to see.

These gibbers in Sturt's Stony Desert are broken-down duricrust, polished by wind-blown sand.

Saltbush, bluebush, spinifex and other plants may grow sparsely[G] on some stony deserts.

When air is heated over gibbers, a mirage[G] may form. It may show an image of a distant object or of the sky.

Large animals find it difficult to survive on stony deserts unless recent rain has sprouted plants and left waterholes.

Sandy deserts

CREATURES OF THE STONY PLAINS

This grasshopper uses camouflage to escape reptiles, birds and other animals that would like to eat it.

The Gibber Chat is an insect-eater that lives only on dry, stony plains. It uses a rock as a perch.

Dragon lizards shelter from heat or cold under rocks. They hunt small creatures across the stones and bare ground.

The rat-sized Kowari shelters in burrows. At night it hunts creatures up to the size of a mouse.

LURING THE ENEMY

The insect-eating Australian Pratincole (left) breeds on the desert plains in springtime. It nests on the bare ground and its chick (right) can run almost as soon as it hatches. If a predator threatens, the chick squats down, looking like a stone. The adults try to lure the danger away from the chick by lurching along the ground, pretending to be injured.

Desert coasts and islands

Deserts near the sea may be formed in several ways.

They can occur on the western sides of continents where the winds travel over cold ocean currents and do not pick up moisture. Western Australia has coastal deserts of this type in the north-west and along the south coast into South Australia.

Local coastal deserts can take the form of sand dunes along otherwise normal coastlines. The sand islands of south-eastern Queensland, especially Fraser Island, contain many mini-deserts. These are sand dunes that have no anchoring plants.

Newly formed islands such as coral cays may be deserts. Often they have no surface water and the plants that survive on them will grow in salty and dry conditions.

A desert on the coast of south-western WA. Sand dunes border pillars formed when limestone replaced the roots and stems of dune plants.

Unless plants anchor a dune, it will shift as the wind blows its sand away.

On the Peron Peninsula, Shark Bay, WA, white sea sand and red desert sand border the ocean. The salt lakes are cut-off sea lagoons.

Bare areas on sand islands such as Fraser Island, Q, are mini-deserts.

Desert coasts and islands

COASTAL DESERT PLANTS

Fleshy leaved pigface grows around salt lakes and in coastal dunes.

Goats-foot Convolvulus creeps across coastal sand dunes.

Coastal grasses send out runners to form new tussocks.

Coastal desert plants grow low and have leaves that save water.

COASTAL BIRDS

The Pied Oystercatcher feeds along the tide line and on sandflats.

The Red-tailed Tropicbird fishes at sea but nests on island shores.

DESERT ISLANDS

Islands where fresh water is scarce or absent are like deserts. The plants and animals that live there have to survive dry, salty conditions. Birds get rid of salt through glands[G] opening into their nostrils. Plants may store salt in their leaves, then shed the leaves.

A coral cay has habitats that are dry and salty.

COASTAL DESERT ANIMALS

Hermit crabs shelter in burrows above high-water line.

Ghost Crabs patrol the shoreline looking for food.

The Silver Gull is a scavenger. It may nest on desert lakes.

Water in the desert

Water may come to a desert in the form of rain. Much of it evaporates; some soaks into the ground. Here it is taken up by plants or remains as groundwater. If the groundwater meets a layer of impermeable[G] rock, it may filter towards lower ground through porous rock and soil. This means that underground, or artesian, water may come out on the surface hundreds, or even thousands, of kilometres from where it fell as rain.

Surface run-off may flood the desert. Sometimes, if the rain has been heavy enough, it hurtles down dry watercourses as a flash flood. It drains away into channels and creeks. These flow into rivers that may travel long distances to end in inland salt lakes. Desert rivers flow across mainly flat ground and are usually winding.

A desert river filled by recent rain. The cut-off pools are called billabongs.

Desert rain may fall on a small local area. This heavy fall is at Uluru. The waterfalls show where water has worn away the rock.

A waterhole in an otherwise dry riverbed becomes a source of water for many plants and animals.

Kangaroos will drink daily if water is nearby.

Desert birds have to watch for predators as they drink.

Water in the desert

FOUND NEAR WATER

The Zebra Finch eats seed.

A Willie Wagtail with dragonfly prey.

A Spotted Harrier at its nest.

A mud-dauber wasp and prey.

Many creatures are found near desert water sources. Seed-eaters such as finches and pigeons drink each day. Large animals, such as kangaroos, also need to drink. Some birds and insects need mud to build their nests. In a dry season, water sources shrink and many thirsty creatures visit them. Predators hunt the drinking animals. These predators often breed near water, where they can get food for their young.

The Magpie-lark is a mud-nester.

ARTIFICIAL WATER SOURCES

Corellas flocking to water in a trough supplied by a windmill.

A crow and Apostlebirds look for water at a tap.

Humans have made water available in the desert. They pump up underground water from the Great Artesian Basin and irrigate dry land by running pipelines from rivers and dams. These sources are used by wildlife. However, water resources are not endless. Over-use is draining the Great Artesian Basin. Taking water from rivers for irrigation means salt levels rise and rivers become polluted[G].

Plagues, poverty, plenty

Desert plants and animals must be able to breed up quickly once conditions are right.

After enough rain falls, desert plants grow new leaves. Then they flower and produce seeds. Land creatures such as insects and water animals such as shrimps feed on the new growth and on each other, and breed in great numbers. Animals that eat plants, their nectar and their seeds court and mate, and have young ones.

Creatures such as grasshoppers, mice and rats appear in great numbers. Where they appear on farming or grazing land, they are said to be in plagues. Carnivores hunt the herbivores and scavengers eat whatever they can find.

When the water dries up, many plants and animals die. However, they leave seeds and young ones behind to survive until the rain comes again and they can breed in their turn.

BUDGIE BLISS

After a breeding season, Budgerigars gather in flocks. They roam the arid grasslands, feeding on grass seeds.

After rain falls, Budgerigars pair up and look for nest hollows. They may bring up several broods.

THE SAGA OF THE STILTS

Banded Stilts usually live on coastal salt lakes, where they eat tiny brine shrimps. After good rains, brine shrimps appear in huge numbers in inland salt lakes. The stilts fly to these faraway lakes to breed. They will nest until the lakes dry up, then fly back to the coast with their young.

Stilts on an inland lake.

MYSTERY MOUSE

Plains Mouse and young ones.

The Plains Mouse is now found only on the gibber plains of the Lake Eyre basin. It lives in colonies in burrows connected by surface runways. No one knows why it is disappearing.

Plagues, poverty, plenty

A night-hunting Letter-winged Kite with a Long-haired Rat.

This Letter-winged Kite has brought a Long-haired Rat to its chicks. When rain falls on the gibber plains of western Queensland, these rats breed in huge numbers. The kites hunt from dusk to dawn to feed their chicks. The female may be sitting on a second clutch of eggs while the male brings up the first brood. As rats disappear, the chicks are fed on lizards. Then the kites give up breeding until the next rat "plague".

FROGS A'WOOING

Frogs mating. They use their "spade feet" to dig themselves in.

Many desert frogs sleep away dry weather buried deep in the ground. When it rains, they surface and look for food. The males call and mate with the females that answer them. Desert tadpoles rapidly change into frogs. As the water dries up, they dig themselves into the ground to await the next rain.

THE DYING TIME

When the desert dries out, butterflies may drift to other areas on the wind.

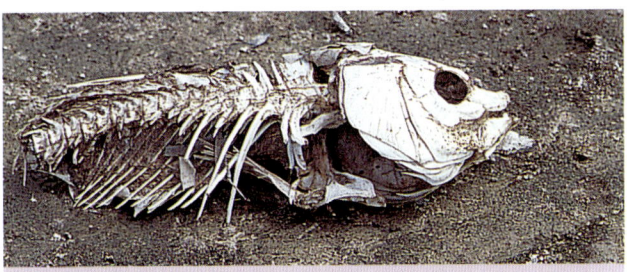

Fish are trapped and die in drying waterholes.

Young birds that are unable to fly will die when the water dries up and their parents leave.

Some creatures can escape from the desert when the time of plenty is over. Others are trapped and die, leaving their bones on the drying ground.

Ferals in the desert

▶ **A feral animal is one that at some stage in its history was domesticated or tamed by humans, but has gone wild.**

The earliest large feral animal in Australia was probably the Dingo, which came with seafarers from South-east Asia around 3500 years ago. The feral cat was possibly brought to Australia by seafarers. When Europeans arrived, they brought sheep, cattle, pigs, goats and horses as well as more cats and dogs. Black and brown rats and the house mouse came as vermin[G] on their ships.

One-humped camels arrived from 1840. The European rabbit was introduced in 1858, the brown hare in the 1830s, the fox in the 1860s and the donkey was first imported in 1866.

Feral grazing animals and domestic stock eat young plants, destroy desert waterholes and damage the soil. However, a farmer can move domestic stock off an overgrazed area. Feral grazers have to be destroyed.

The fox and cat are predators. Re-establishing native animals in an area depends on their removal.

THE HERBIVORES

The feral goat competes with native animals for food and shelter. It may also carry foot and mouth disease.

Many people become very worried at the thought of getting rid of feral animals, particularly feral cats and horses. Feral animals are wild animals. Some, like cats and foxes, could never be made into pets. Others, such as camels and brumbies, can be caught and tamed with difficulty. However, few people would have the facilities and knowledge to use them. They destroy habitat, compete with native animals and can carry diseases.

The way in which sharp hooves cut up the soil can be seen in this picture of desert wild horses, or brumbies, running.

A camel's soft foot-pads cause less harm to the soil than the sharp hooves of cattle, sheep, horses and donkeys.

Ferals in the desert

THE PREDATORS

A feral cat on the rocky slopes of a desert range.

A fox prowls across a clearing in bluebush country.

WEEDS

Plants that have become pests, or weeds, have a number of things in common.

• They are attractive to humans for some reason, e.g. as shade plants or stock feed. This accounts for their original introduction.

• They can grow in a wide range of habitats, survive stress well and grow fast. This establishes them in an area.

• They reproduce easily. Their seeds spread widely and germinate^G easily. They may send out runners, or use other means of spreading that do not involve seeds. This lets them take over new areas quickly.

Noxious weeds are plants that actively harm habitats and are seen as harmful by humans.

Prickly acacia, mesquite and buffel grass are major plant pests in Australia's semi-arid grasslands. They crowd out and overshade native plants.

A pig using its snout to dig up the roots of water plants. Weeds will take over the places where it destroyed native plants.

FACTS 'N' FIGURES FILE

• **The camel eats** more than 80% of the plant species in its environment. It prefers those with high water and salt content, but eats thorny, bitter plants other mammals will not eat.

• **A thirsty camel** may drink up to 200 L of water in 3 min. In winter it can go without water for several months if its food plants are moist enough.

• **Feral goats** grew in numbers after Dingos were shot and trapped to protect sheep. Goats would rather browse trees and shrubs than eat grass. They will climb trees to reach the leaves.

• **The rabbit eats** seedling plants. The goat eats bushes. Between them they can strip a desert area of vegetation.

• **A single fox can kill** an animal up to 5 kg in weight (such as a small wallaby). A cat can kill an animal up to 2 kg in weight.

• **Cats may have arrived** in Australia from Dutch shipwrecks on the coast of Western Australia, or with Asian fishermen.

• **Rabbits build up quickly** in good seasons and strip the ground of grass and seedling plants. The rabbit calicivirus, which has been introduced to control rabbit numbers, works better in desert areas than in wetter regions.

• **A herd of feral goats** can increase by 75% each year.

• **Donkeys** were introduced to northern Australia to replace horses, which fell victim to walkabout disease (see p. 13).

People and deserts

Australia's soils are the poorest of those found on any continent.

They were made long ago and, since there is little earth movement in the form of quakes or volcanos to bring new rocks to the surface, they are not being renewed. The surface of Australia's deserts is not protected by thick vegetation. Landscapes are easily eroded, and once the surface soil is gone, the plants and animals go as well.

Many tourists visit the desert and there are people who live and work in deserts. Everyone needs to become aware of the harm that domestic stock and human activities, such as four-wheel-driving, can do to the fragile desert surface.

Desert ecosystems are easily damaged and must be protected.

DESERTS NEED PROTECTION

Desert landmarks, such as Uluṟu and the nearby Kata Tjuṯa, attract thousands of tourists each year. Unless these visitors take care, they will harm the fragile desert environment.

The Devils Marbles are a desert landmark. They are beside a major road and need to be protected from vandals^G.

Camels are still used to carry people through the desert. They harm the landscape less than most imported animals.

People and deserts

ABORIGINAL DESERT-DWELLERS

An Aboriginal girl eating a quondong fruit.

Making a spear from desert acacia wood.

The Aboriginal people have lived successfully in Australia's deserts for tens of thousands of years. They were nomads who moved to places where there was food and water. They burned areas to encourage new plant growth that attracted animals they could hunt for food.

LEARNING ABOUT THE DESERT

Desert education centres in places such as Alice Springs, NT, let people see desert plants and animals close up.

WORLD HERITAGE DESERT SITES

At present (September 2001) Australia has 14 places on the World Heritage List. Four of them are in deserts. They are:

- The Willandra Lakes, NSW
- Shark Bay, WA
- The Australian Fossil Mammal Site, Riversleigh, Q
- Uluru–Kata Tjuta National Park.

In addition, Fraser Island, Q, has important sand-dune habitats.

◯ MAKING A LIVING FROM THE DESERT ◯

Some of Australia's largest mining towns are in the desert. This is Mt Isa, in the west of Queensland.

Many people have made homes in Australia's deserts. They make a living from mining, or from grazing sheep or cattle. Tourism is a growing industry.

A ruin marks a failed attempt at farming.

Life underground in an opal-mining town.

In good rainfall years, cattle and sheep do well in the desert. In drought years, many will not survive.

Glossary

Adapt Change to suit new conditions.

Bacteria Tiny life forms.

Camouflaged Coloured, shaped or textured so that a thing is hard to see against its background.

Canopy Leafy branches of trees.

Capsules Small cases.

Cellulose Fibrous material in cell walls of plants.

Claypan A clay-lined hollow in the ground. It can hold water.

Collected Added to a scientific collection.

Colony A group of animals or plants living together.

Community A group of plants and animals living in the same area and interacting with each other.

Crater A cup-shaped hollow in the top of a mound.

Cycle Things that repeat themselves in the same order and at the same intervals.

Dominant Standing above the rest.

Elements Simple substances that cannot be divided by chemical methods into any further parts.

Embryo An organism in the early stages of development.

Endangered In danger of disappearing forever.

Ephemeral (of plants) Growing, flowering and seeding quickly after good rain falls in dry climates.

Erosion The process by which the surface of the Earth is worn away.

Evaporate To disappear as vapour.

Extinct No living examples still exist.

Feral Gone wild.

Fertilised A female cell has been made capable of developing into a new individual by being joined with a male cell.

Fertiliser Of plants, something used to enrich the soil.

Flora Plant life.

Germinate To sprout; to begin growing.

Gibber Stone or boulder.

Glands Organs or tissues that secrete substances.

Home range The area in which an animal finds food, water, shelter and mates.

Impermeable Not allowing liquid to pass through.

Larvae The grubs or caterpillars of insects that have several stages in their life cycles.

Mammals Animals with backbones whose bodies create their own heat and that have hairy skins and feed their babies on milk.

Marsupials Mammals whose young are born at a very early stage of development. The young shelter in a pouch or fold of skin until they grow big enough to live outside it.

Micro-climate The climate of a tiny or confined area.

Minerals Inorganic substances obtained by mining.

Mirage An optical illusion in which distant objects are reflected by a layer of hot air.

Mobile Moving readily.

Mucus A thick liquid secreted by the body.

Native Belonging to the country.

Nectar The sweet liquid produced by flowers.

Nutrients Things used as food.

Ovum The female reproductive cell of plants and animals.

Paralysed Made unable to move.

Parasitic Of organisms, those that live on or in the bodies of other species (hosts), taking nourishment and causing harm.

Pollen Fine yellow grains that fertilise the female cells of plants.

Polluted Made unclean, foul or unhealthy (usually done to the environment).

Porous Letting water pass through.

Predator (*noun* **predation**) An organism that preys upon (kills and eats) other organisms.

Protein A substance based on nitrogen, found in all organic bodies.

Python A snake without poison glands that kills its prey by squeezing.

Reptiles Animals with backbones whose bodies take the temperature of their surroundings and whose skins are covered with scales.

Reproduce To produce a new generation of individuals.

Silica A mineral that is found as sand, quartz, flint and agate.

Soar To fly on rising currents of air without moving the wings.

Sodium bicarbonate A white powder used in cooking, medicine, etc.

Solar radiation Rays that come from the Sun.

Sparsely Thinly scattered.

Steppes Wide, often treeless plains.

Torpid The condition of an animal that has slowed down body activity in order to survive a cold period.

Transpire To lose water vapour from leaves or stems.

Tubers Underground stems bearing tiny leaf buds from which new plants may grow, c.g., a potato.

Unique Of which there is only one.

Vandals People who deliberately damage things.

Vapour Something in the form of gas; sometimes visible, such as fog, mist, steam, smoke.

Vermin Troublesome animals.

Index to subjects illustrated

acacia 8, 20
  wood 45
ants 5, 30
  honeypot 29
Apostlebird 39
Avocet, Red-necked 32

Bat, Ghost 25
beetle, jewel 19
Bettong, Burrowing 29
Bilby 5, 29
billabong 38
bindiis 30
Birdflower, Green 13
bluebush 8, 34
Bowerbird, Spotted 21
breakaway 24
brumbies 42
Budgerigar 40
Bustard, Australian 27
Butcherbird, Pied 15
butterfly 41
Buzzard, Black-breasted 27
butte 24

camel 6, 42, 44
cat 43
caterpillar (as prey) 39
cattle mustering 4
central Australia 3, 7
Chat, Crimson 21
  Gibber 35
chenopod 8, 30, 31
claypans 4, 8, 33
Cockatoo, Major Mitchell 27
Convolvulus, Goats-foot 37
coral cay 37
corellas 39
Crab, Ghost 37
  hermit 37
crow 39
Cuckoo-shrike, Ground 20

Devil, Thorny 5, 29
Desert, Sturt's Stony 34
desert, Aboriginal lifestyle in 13,
    21, 45
  coastal 8, 36
  education centres 45
  environmental protection of 44
  farming and grazing 4, 18, 26,
    39, 45
  mining centres 45
  peas 21
  sandy 6, 8
  stony 8, 34, 35
  United Arab Emirates 6
Devils Marbles 44
Dingo 14, 28

Dragon lizard 35
  Bearded 21
dragonfly (as prey) 39
Dunnart Fat-tailed 16, 27

Eagle, Wedge-tailed 14, 25
Echidna, Short-beaked 19
Emu 26
emu bush 21
ephemeral wetlands 8, 32, 33
erosion, human causes 7
  sand 7
  temperature change 7
  water 7, 38
eucalypts 8, 10, 11, 18
Euro (Common Wallaroo) 5, 25
Everlasting Daisies 21

Fairy-wren, Splendid 17
Falcon, Black 31
ferals:
  camel 42
  cat 43
  fox 43
  goat 42
  horse 42
  pig 43
Finch, Zebra 39
firestick farming 13
fish, skeletal 41
fox 19, 43
Fowl, Mallee 19
Fraser Island 36
Frog, Water-holding 16
frogs mating 41

Galah 19
Gecko 14
  Marbled Velvet 25
  Three-lined Knob-tail 5, 15
gibber plains 8, 34
goat 42
gorge 7, 8, 13, 24
grasses 5, 8, 11, 12
  coastal 37
grasshopper 35
grasslands 4
  hummock 8
  tussock 8, 26
grazing 4, 26
Grebe, Australasian 33
Grevillea, Honey 13
Gull, Silver 37

Hare-wallaby, Rufous (Mala) 23
Harrier, Spotted 39
herbs 10
honeyeater 38
  Singing 19

Hopping-mouse, Spinifex 16, 23
horse 42

kangaroo 38
  Grey 19
  Red 14, 17, 26, 31, 33
Kestrel, Australian 27
Kite, Letter-winged 41
Kowari 35

lakes, salt 8, 32, 36
Lizard, Centralian Blue-tongued 23
  legless 14

MacDonnell Ranges 24
Magpie-lark 39
Mala (Rufous Hare-wallaby) 23
mallee country 8, 18
Marshwort, wavy 13
mesa 7
minirichie tree 21
mistletoe 11
Mole, Marsupial 29
Monitor, Ridge-tailed 25
  Sand 15, 19
Mt Isa 45
mountains, rain shadow 6
Mouse, Plains 40
mulga country 8, 20
  wood 21
mulla mulla bush 21

Nullarbor Plain 30

Oystercatcher, Pied 37
Owl, Grass 27

Palm, Red Cabbage 13
Parakeelya 29
Parrot, Hooded 27
peaflowers 21
'Peewee' (Magpie-lark) 39
Pelican, Australian 17
  (skeleton) 41
Peron Peninsula 36
pig 43
Pigeon, Spinifex 23
pigface 37
planigale 27
plant adaptations 11
Pratincole, Australian 35
poverty bush 21
Python, Woma 5

Quondong 12, 45

ranges 6, 8
  MacDonnell 24
Rat, Greater Stick-nest 31
  Long-haired (as prey) 41
Ringneck, Mallee 19

river bed 7, 24, 38
Rock-wallaby, Black-footed 15
  Yellow-footed 25
rocky landforms 5, 7, 8, 10, 24,
    34, 35
Rose, Sturt's Desert 29

salination 32
saltbush 8, 30, 34
saltpan 32
samphire 32
sand dunes 4, 8, 12, 28, 33,
    36, 37
  plains 8, 28
Shark Bay 36
she-oak (casuarina) 11
Shingleback 19
shrublands, acacia 8
  chenopod 30
  eucalypt 8, 18
shrubs 8, 10
skink 14, 17, 19
spider 29
spinifex 5, 8, 11, 14, 22, 34
springs 24
steppes, shrub 8
Stilt, Banded 40
stone country 5

termite mound 5, 14, 27
tourism 44, 45
tracks 28
trees 10
Tropicbird, Red-tailed 37

Uluru 3, 38, 44

wadi 7
Wagtail, Willie 39
Wallaby, Bridled Nailtail 3, 15, 23
Wallaroo, Common (Euro) 5, 25
wasp, hunter 29
  mud-dauber 39
water:
  artificial sources 39
  fall 38
  hole 38
  tap 39
  temporary sources 8, 32, 33,
    38, 40
  trough 39
Whipsnake, Yellow-faced 23
wildflowers 2, 3, 12, 13, 20, 21,
    29
witchetty grubs 21
Wombat, Southern
  Hairy-nosed 15
Woodswallow, White-browed 16

Some information sources

Websites: A few websites that are full of interesting information are listed here. There are many more. Just type into your search engine some key words, such as "australia deserts", "australia salt lakes" or "australia desert animals".

All Australia, LinkAustralia: www.allaustralianwebsite.com

Australian Broadcasting Commission (ABC), natural history: www.abc.net.au/abcinternational/catnath

Australian Museum: www.austmus.gov.au

Australian Surveying and Land Information Group (AUSLIG):   www.auslig.gov.au

Department of Conservation and Land Management (CALM), Western Australia: www.calm.wa.gov.au

Environment Australia: www.ea.gov.au

The Wilderness Society: www.wilderness.org.au

Books: Search for these books in your local or school library. Look in the contents and index, check the subjects in which you are interested, then find the pages on which they appear. There are many more books on Australian deserts and the habitats they contain. Also check the newspapers for current-events information.

Beattie, Andrew J. (Ed.), 1995. *Australia's Biodiversity: Living Wealth*. Reed New Holland, Sydney.

Cogger, Harold G., 1999. *Reptiles and Amphibians of Australia*. Reed New Holland, Sydney.

Greig, Denise, 1999. *Field Guide to Australian Wildflowers*. Reed New Holland, Sydney.

Holiday, Ivan, 1989. *A Field Guide to Australian Trees*. Reed New Holland. Sydney.

Morcombe, Michael, 2000. *Field Guide to Australian Birds*. Steve Parish Publishing, Brisbane.

Recher, Lunney & Dunn (Eds), 1992. *A Natural Legacy: Ecology in Australia*. Maxwell Macmillan Publishing Australia, Sydney.

Reid, Ian G., 1994. *The Bush: A Guide to the Vegetated Landscapes of Australia*. UNSW Press, Sydney.

Slater, Peter, 2000. *The Slater Field Guide to Australian Birds*. Reed New Holland, Sydney.

Smith, David, 1994. *Saving a Continent*. UNSW Press, Sydney.

Stanton, J., 1995. *The Red Centre*. The Australian Geographic Society, Sydney.

Strahan, Ronald (Ed.), 1995. *The Mammals of Australia*. 2nd ed. Reed New Holland, Sydney.

White, Mary E. (Ed.), 2000. *Running Down: Water in a Changing Land*. Kangaroo Press, Sydney.

Van Oosterzee, Penny, 1998. (Ed.), *A Field Guide to Central Australia*. JB Books, South Australia.

First published in Australia by Steve Parish Publishing Pty Ltd
PO Box 1058, Archerfield, Queensland 4108 Australia

www.steveparish.com.au

© Copyright Steve Parish Publishing Pty Ltd

ISBN 1740210891

Photography: Steve Parish

Text by Pat Slater, SPP
Designed by Leanne Staff, SPP
Edited by Kate Lovett, SPP

Additional photography: Stanley Breeden – ant, Woma p. 5, mistletoe p. 11, quondong p. 12, grassfire p. 13, monitor p. 15, woman digging, witchetty grubs p. 21, monitor p. 25, mole, honeypot ants p. 29, Kowari p. 35, water on Uluru p. 38, frogs p. 41, girl, making a spear p. 45; Pat Slater – camel train p. 6; Peter Slater – succulent p. 11, woodswallow p. 16, fairy-wren p. 17, cuckoo-shrike p. 20, chat, minirichie, bowerbird p. 21, pigeons p. 23, kestrel p. 27, hunter wasp p. 29, falcon p. 31, avocet p. 32, grasshopper, chat, lizard, pratincole & its chick p. 35, honeyeater p. 38, finch, Willie Wagtail, harrier p. 39, kite, butterfly p. 41; Raoul Slater – parrot p. 27, grebe p. 33, mud-dauber wasp p. 39.

Film by Inprint Pty Ltd, Brisbane, Queensland
Printed in Singapore by Craft Print International

Designed and produced in Australia at the Steve Parish Publishing Studios